Museum of Science

OFFICIAL COMMEMORATIVE GUIDE

Larry J. Ralph and Mary Ann Trulli

BECKON BOOKS

ON THE WATER
One of the world's largest science centers, the Museum of Science is a prominent Boston landmark and one of Boston's most attended cultural institutions, hosting about 1.5 million visitors a year.

BANNER UP
The tower of the Central Building is adorned with banners highlighting the Museum's temporary exhibits and Mugar Omni Theater shows.

Building a Science Museum

In 1830, six men dedicated to the advancement of natural history established the Boston Society of Natural History—a place where they could discuss their research and share their collections. The society moved to its first permanent building in 1864, and the new Museum was named the New England Museum of Natural History. In 1939, famed explorer, cartographer, and aerial photographer Bradford Washburn became the Museum's director, transforming it from a collection-based repository of natural history specimens into a world-renowned science education institution. Washburn negotiated a 99-year lease for land spanning the Charles River and gave the Museum a new mission: to feature all the sciences under one roof. The renamed Boston Museum of Science—today called the Museum of Science, Boston—officially opened its new building in 1951.

Over the years, the Museum added many popular exhibits and initiatives, including the Charles Hayden Planetarium in 1958, the Elihu Thomson Theater of Electricity in 1980, the Mugar Omni Theater in 1987, and the *Gordon Current Science & Technology Center* in 2001. In 1999, the Museum combined forces with The Computer Museum in Boston, bringing new visitor experiences to the Museum's programs. In 2004, the Museum launched the National Center for Technological Literacy, which aims to integrate engineering into school curricula nationwide. Together, these initiatives further the Museum of Science's mission to transform the nation's relationship with science and technology.

MUSEUM MASCOT
Found abandoned in the woods in 1951, Spooky, a great horned owlet, was only a few days old when he was offered into the Museum's care. Raised by Museum staff and at ease with people, Spooky soon became a major celebrity. He was photographed for *Look* magazine and made TV appearances on both CBS and NBC, including the *Today* show.

MILESTONES
Spooky the owl was the star of many live presentations at the Museum and became one of its most memorable icons. By the time he passed away at age 38, he was the oldest known member of his species.

IMPRESSIVE DISPLAY
The New England Museum of Natural History, shown in 1930, was the forerunner of the Museum of Science. The main exhibit gallery featured a 48-foot skeleton of a finback whale, given to the Museum in the 1870s, that was suspended overhead. The skeleton was surrounded by hundreds of display cases filled with natural history specimens.

Building a Science Museum

DRIVING MR. DINO
The head of the Museum's first full-scale *T. rex* model, which now stands in front of the building, was paraded down the Massachusetts Turnpike in 1966 to help raise money to build the West Wing (now the Blue Wing).

MILESTONES
In April 1949, over a thousand small fish "rained" down on the site of the Museum. This phenomenon may have been triggered by strong downdrafts disturbing the flight of seagulls with full stomachs.

DEVELOPING THE DOME
Construction on the Charles Hayden Planetarium, top, began in 1952. The Planetarium opened in 1958 and immediately became a visitor favorite with its spectacular excursions into the night sky.

GREAT HEIGHTS
The Museum of Science founding director, Bradford Washburn, demonstrates his mountain climbing skills in 1963 as he rappels down the Central Tower to inspect newly installed sculptures.

MUGAR OMNI THEATER
In the Mugar Omni Theater, the world's largest film format is projected onto a five-story-tall IMAX Dome that wraps around the audience and immerses visitors in the action.

WHAT'S SO BIG ABOUT THE IMAX FILMS SHOWN IN THE MUGAR OMNI THEATER?

The 45-minute feature films shown in the IMAX Dome theater are almost three miles long and weigh about 200 pounds on the spool. The film format used in IMAX projectors is 10 times larger than the traditional 35mm film used in Hollywood productions.

T. Rex on Duty
The Museum's first full-scale *Tyrannosaurus rex* model, built in the 1960s, stood erect. When a more current model was created for the dinosaur gallery in 2001, the Museum gave the old fellow a new home outside the Mugar Omni Theater.

MILESTONES

During construction of the main building between 1949 and 1951, the Museum of Science was temporarily housed in a small pavilion still seen in the park behind the Museum today.

Outer Spaces

Situated on seven acres that span the Charles River, the Museum of Science is a regional landmark. The building was constructed on a dam designed to contain the river basin and stabilize it from tidal fluctuations. Several of the Museum's displays are outside the building. On the roof are several different styles of small wind turbines that are part of the *Catching the Wind* display in the Exhibit Halls, as well as a large photovoltaic array that converts sunlight into electricity. In addition, five gold-leafed aluminum sculptures are mounted along the riverfront façade of the building. Commissioned in the early 1960s, these figures represent the Museum's five principal program areas at the time: Astronomy, Man, Nature, Energy, and Industry.

Near the Museum's entrance is the *Rock Garden*. This exhibit highlights 21 large geological specimens from around the globe, gathered from places as diverse as Alaska's Mount McKinley and the bedrock under the Museum. Visitors are encouraged to touch the specimens, which include petrified wood from Arizona, granite from Egypt, a two-ton chunk of limestone from the Rock of Gibraltar, and Boston's own Roxbury puddingstone.

ROCK STARS
Rock formations from around the world are displayed in the Museum's *Rock Garden*, above.

MEANINGFUL SCULPTURES
The five large sculptures located on the rear face of the Museum were commissioned in 1962. Created by Theodore C. Barbarossa, they depict the five major Museum programs at the time: Astronomy, Industry, Man, Nature, and Energy.

HOW DOES WOOD BECOME PETRIFIED?

Most trees decompose after they die, but if the conditions are just right, they can be transformed into rock. The mineral silica, found naturally in the soil and dissolved in rainwater, slowly replaces the cell walls of the wood and then crystallizes into quartz. The wood is gone, but the tree's shape and patterns remain in the rock.

New England Habitats

In *New England Habitats*, a range of nature environments are brought to life in six historically classic dioramas: the Katahdin Woods of central Maine; Mount Desert Island, along the coast of Maine; the Green Mountains of Vermont; Squam Lake and Wildcat Mountain in New Hampshire; and Crane Beach, on the coast of Massachusetts.

This exhibit is designed to appeal to the five senses. Visitors can touch models of birds and casts of feet, antlers, and beaks; hear the sounds of a beaver's tail smacking the water in alarm; or breathe in the scents of a spruce forest or seashore. Other displays compare the features of various animals with human-made tools. These reveal how the adaptations of New England's wildlife allow the animals to thrive in diverse, sometimes harsh, environments.

BUSY BEAVERS
A small pond and stream are home to a family of beavers in this diorama.

ANIMAL INSPIRATION
Humans may have copied animals for many important inventions. For instance, human-made tools such as pliers and chisels work much like pointed bird beaks and beaver teeth.

OUR NATURAL WORLD

Beavers spend the winter swimming in ice water and have thick, fuzzy undercoats that trap air, acting as insulators and slowing down the loss of heat from their bodies. Their fur also has a coating of oil that makes it waterproof.

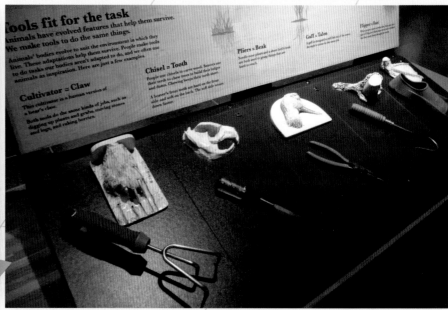

WHAT'S THE DIFFERENCE BETWEEN SLEEP AND HIBERNATION?

Hibernating animals seem barely alive. They feel cold to the touch, their heartbeats and breathing are nearly imperceptible, and they are very hard to awaken. In contrast, when animals are asleep for the winter, their heartbeats and breathing are close to normal, and they can be awakened easily.

FALL FORAGE
A black bear mother and her cubs forage for food before their long winter sleep. Snow-covered Mount Washington is shown in the background.

MONKEY BUSINESS
The Museum's tamarin monkeys are small, squirrel-like monkeys raised as part of the Association of Zoos and Aquariums' Species Survival Plan (SSP). The SSP helps ensure the preservation and breeding of many endangered wildlife species.

OUR NATURAL WORLD

Animals come in a variety of shapes and sizes, but they all have two things in common: they are multicellular (made of many cells), and unlike plants, they cannot make their own food. They must get nutrients by eating other living things.

Live Animals

The Museum of Science was the first science and technology center in the U.S. accredited by the Association of Zoos and Aquariums. The Museum cares for over 100 animals representing more than 50 species in its Live Animal Center. Although the Center itself is closed to the public, visitors can see a selection of animals from the viewing windows and watch how staff members care for them.

Some of these animals are stars in the Museum's live animal presentations on the Shapiro Family Science Live! Stage, while others are featured within various exhibits. There are camouflaged reptiles and amphibians in *Take a Closer Look*, butterflies and other insects in the *Butterfly Garden*, and a variety of small species in the *Discovery Center*. In other areas of the Museum, cotton-top tamarin monkeys provide a glimpse of the social behavior of family groups, and the chick hatchery reveals the wonder of birth. One of the most popular exhibits featuring live animals is the glass-enclosed beehive. There, thousands of bees travel in and out of the colony, collecting pollen, making wax, capping honey, and tending to the brood while the queen bee lays her eggs.

UP CLOSE AND PERSONAL
An Aldabra tortoise from Aldabra Atol, Africa, above, takes part in an animal demonstration on the Science Live! Stage.

BUZZING AROUND
Both children and adults enjoy the challenge of finding the queen bee, who is often concealed among hordes of worker bees as she lays her eggs.

MILESTONES
The Museum's 42-pound lobster held the world record for over 40 years—from its capture in 1934 until 1977, when a 44-pounder was caught off the coast of Nova Scotia.

Live Animals

TROPICAL REPTILE
The prehensile-tailed skink is native to the Solomon Islands in the southwest Pacific. Active mainly at dusk and dawn, this skink makes its home in the trees, using its curled tail to hold onto branches and green skin to hide among the leaves. This plant-eating lizard is the largest of the skinks, growing to over 30 inches long.

BURROW MAKER
The woodchuck, or groundhog, is common to the northeast United States and is an excellent burrow maker, creating complex underground tunnel systems.

ON THE HUNT
Eastern screech owls typically nest in hollowed-out tree trunks. Like all owls, they have sharp claws well suited for hunting small prey.

MILESTONES

Between January 1947 and the new building's opening in February 1950, the Museum took the show on the road in a station wagon packed with science props and live animals. Thousands of children saw these shows, which featured a flying squirrel, a crow, turtles, and snakes.

CURIOUS CRITTER
Long and slender, ferrets are inquisitive by nature. These playful mammals are related to minks and weasels, and are known for their ability to squeeze through the tightest of spaces.

SCALY CLIMBER
Green tree pythons are excellent climbers. They spend much of their time in trees but make their diet of small mammals, not birds.

SPREADING THEIR WINGS
Great horned owls can grow to more than 20 inches tall and have wingspans of over four feet.

Butterfly Garden

In the *Butterfly Garden*, visitors can get close to a variety of live butterflies and moths that fly freely amid the exotic plants. The conservatory highlights what makes each species unique—such as size, color pattern, and wing shape—and describes how the butterfly's colors and wing patterns act as camouflage, mimicry, or a warning signal to predators. Printed guides with photographs identify the butterflies and moths, and labels among the foliage highlight key behaviors such as feeding, courting, and basking.

Other displays and activities demonstrate the four different stages of the butterfly's lifecycle: egg, caterpillar, chrysalis, and adult. An "emergence box," where hanging chrysalides transform into adult butterflies, shows the process of metamorphosis. If the timing is right, visitors might see a new butterfly crawling out of its casing or a recently emerged adult pumping fluid into its wings.

The Garden also explores the butterfly's relationship to its animal relatives and ancestors. Inside various terrariums, there are live specimens like giant centipedes, cockroaches, and tarantulas.

ZEBRA STRIPES
The zebra longwing butterfly is native to Florida and Central America.

INSECT OBSERVATION
Visitors can get close to dozens of species of butterflies and other insects in the *Butterfly Garden*.

OUR NATURAL WORLD

Most butterflies have pigments in their cells that absorb light. The wavelengths that aren't absorbed are reflected, and the reflected light is perceived as color.

WATERFRONT VIEWS
The *Butterfly Garden* provides a view of the Boston and Cambridge skylines and the Charles River.

WINGS AND THINGS
The julia butterfly is identified by its bright orange wings, which alert predators that it will make an unpleasant meal. It belongs to the heliconians (longwings) subfamily of butterflies. As caterpillars, many of these butterflies eat passion vines that make them unpalatable to their prey.

ARE HUMANS RELATED TO BUTTERFLIES?

All living things on Earth—not just animals but plants, bacteria, and more—are descended from a common ancestor that lived more than three billion years ago. Just as human cousins share a common grandparent, humans and butterflies share a common ancestor that lived between 600 and 900 million years ago.

Dinosaur Galleries

The fossils that scientists collect tell a still-evolving tale of prehistoric life on Earth. In the dinosaur galleries, the fossils and life-size models on display depict the changing way paleontologists view these extinct animals. For example, the Museum's first *Tyrannosaurus rex* model, built in the 1960s, was based on the only evidence that paleontologists had at the time: three incomplete skeletons. Today, scientists have uncovered more than 30 *T. rex* skeletons—some almost complete. These findings led to the 2001 creation of the Museum's agile and trim *T. rex* model, on display in *Modeling the Mesozoic*.

Modeling the Mesozoic also examines fossil remains to ask what separates dinosaurs from other ancient animals. There are "bone dictionaries" with mystery skeletons as well as murals depicting life in the Mesozoic period. Interactive exhibits highlight new scientific methods that determine the walking pace of a *T. rex* and the geography of the Earth during the age of the dinosaurs.

Nearby, *Colossal Fossil: Triceratops Cliff* showcases a 65-million-year-old *Triceratops* fossil, one of only a few nearly complete *Triceratops* on display in the world. Discovered in the Dakota Badlands in 2004, the 23-foot-long specimen was missing some bones. Model makers in Italy cast replacement bones and mounted the specimen in its fully articulated position. Today, visitors can view the *Triceratops* as it would have appeared roaming the Midwest during the Late Cretaceous period.

FIERCE PREDATOR
At 40 feet long and seven tons, *T. rex* was king of the dinosaurs between 68 and 65 million years ago.

GIGANTIC DINOSAUR
This is a full-scale replica of the skull of a *Giganotosaurus*. Even larger than *T. rex*, this meat-eating dinosaur lived 112 to 90 million years ago in South America.

FISHY FOSSIL
Mene, or moonfish, lived in the Eocene Epoch 56 to 34 million years ago.

OUR NATURAL WORLD

Only a very small percentage of fossils are ever found. In many cases, bones are preserved but are embedded in a layer of sedimentary rock that is buried under a mountain, forest, or sea floor. Skeletal remains are often washed away or torn apart by animals. Dinosaur skeletons are rarely found intact.

Dinosaur Galleries

DINOSAUR SLEUTHS
Once paleontologists discover fossilized bones and carefully remove them, they compare these partial skeletons with those of modern animals.

BACK IN TIME
An interactive computer program takes visitors back to the Early Jurassic period 200 million years ago.

OUR NATURAL WORLD

New species of living things have been both evolving and becoming extinct for about 3.5 billion years. Fossils preserve a record of these changes. Knowing when fossilized species existed can help scientists learn the ages of the rocks around them. This technique is called relative dating.

OCEAN ANCESTORS
These Trilobite fossils, right, are from the Ordovician Period, 480 to 440 million years ago. These marine arthropods are relatives of modern horseshoe crabs.

REMARKABLE FIND
This *Triceratops* skeleton, named Cliff, is one of the few mostly complete *Triceratops* skeletons in the world. Cliff was found in the Dakota region of the United States in a fossil-rich layer formed 65 million years ago called the Hell Creek Formation. This layer marks the end of the dinosaur era.

WHY DID THE *TRICERATOPS* HAVE SUCH A BIG "FRILL" ON THE BACK OF ITS HEAD?

Scientists aren't sure. Some believe that along with its horns, the frill was used as defense, protecting the animal's head and neck. Others think it probably helped regulate the *Triceratops*'s body temperature or was used to attract mates.

MODEL MANIA
The giant grasshopper model and miniature train in front of *Making Models* put into perspective how modeling lets us see things in different ways for a variety of purposes.

REPTILE REMAINS
This fossil, above, is of a *Kelchousaurus*— a reptile that lived in what is now China during the Triassic Period, 250 to 200 million years ago.

TAKING FLIGHT
The *Nyctosaurus*, left, was a flying inhabitant of the Late Cretaceous Period, 85 to 65 million years ago. It had a large crest and a wingspan that may have reached ten feet.

HOW FAST COULD A *TRICERATOPS* RUN?
We don't know! In fact, many paleontologists now think that the *Triceratops* held its front limbs in a sprawled out fashion, which would make it difficult to gallop. This doesn't mean *Triceratops* couldn't move fast, but its top speed remains a mystery.

RARE FOOTPRINT
This cast is of the first *Tyrannosaurus rex* footprint ever found. Discovered in New Mexico, the footprint demonstrates that like other therapods, *T. rex* had three front toes, each with a claw. To date, only one other *T. rex* print has been identified—in the Hell Creek Formation of Montana in 2006.

Science Is an Activity

In this group of exhibits, visitors are invited to think like scientists and learn skills to understand the natural world by exploring four distinct activity centers. One exhibit in this area, *Take a Closer Look*, focuses on observation and illustrates how humans can engage their senses—both naturally and by using technology—to investigate the natural world. *Investigate!* helps visitors think like scientists by asking questions, experimenting, and exploring such phenomena as light, motion, and visual illusions. *Natural Mysteries* highlights the importance of classification and provides tools for classifying a wide variety of items, including rocks, minerals, leaves, shells, skulls, and animal tracks. Finally, *Making Models* helps visitors understand, analyze, and create one of science's most basic tools: models. These exhibits highlight various scientific disciplines, communicating that science is more than learning facts and information—it is a process for better understanding our world.

FORWARD MOTION
For many years, this Pacific Type 1927 locomotive, above, was used primarily for hauling passenger trains.

SENSATIONAL SENSES
Take a Closer Look allows visitors to use their senses to view and explore the world around them. With the help of some unique tools, phenomena that are too fast, too slow, too big, too small, or otherwise invisible are revealed.

COLOSSAL CLAMS
Clams can grow to four feet across and weigh over 500 pounds. Giant specimens are found in the Indian and Pacific Oceans. Contrary to a popular myth, these creatures of the sea will not eat divers that come too close.

OUR NATURAL WORLD

How is smell different from sight and hearing? Light and sound exist in a continuous spectrum of frequencies, some section of which humans can perceive. Smell, however, requires the ability to identify up to 10,000 different molecules. There is no single linear scale for smells.

Science Is an Activity

PLAYGROUND LESSONS
Visitors become part of the experiment as they learn the facts about pendulums in the Swing area in *Science in the Park*.

LIGHT PLAY
In *The Light House* exhibit, visitors can play with bending, bouncing, and beaming light, creating artistic lighting effects using mirrors, lenses, and colored filters.

REFLECT AND REFRACT
Rows of tilted mirrors or prisms that reflect incoming light onto a single spot are termed Fresnel lenses. This array of 72 mirrors has two focal points and is essentially two interlocking Fresnel lenses. For more than a century, Fresnel lenses were used in lighthouses to focus a beam of light far out to sea.

OUR NATURAL WORLD

The universe is a vast ocean of electromagnetic waves traveling at the speed of light. Some waves bounce off people and objects, some pass through, and some pass by them completely, depending on the length and frequency of the wave.

LIGHTHOUSE ART
This kinetic light sculpture by Bill Wainright (1999) tops the model lighthouse and creates a rainbow of light that guides approaching visitors to the light and optics exhibit.

Take a Closer Look

In this exhibit, visitors use their senses of sight, hearing, touch, and smell as tools for discovery. Through interactive activities, *Take a Closer Look* highlights the sensitivity of fingers, the limits of hearing, and the ability to hear a melody. It also explores how technology enables people to reach beyond the capacity of their senses, helping them to perceive things that are too fast, too small, too far away, or are simply invisible to the naked eye. Features include an infrared camera that allows visitors to "see" heat, a cloud chamber that detects subatomic particles that penetrate the atmosphere, and a device that uses sand patterns to visualize sound waves.

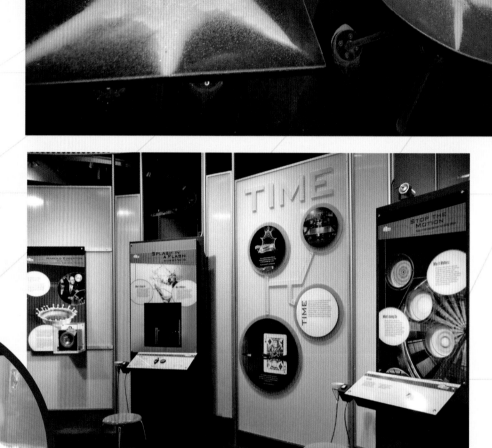

WHITE HOT
Infrared cameras like this one are sensitive to heat sources in the thermal infrared range and display them as visible light. Hot areas appear yellow or white; cool areas appear blue.

SANDY SCIENCE
Chladni plates, top, display patterns in sand resulting from the resonating of metal plates when stroked with a bow. Sand collects at the points of least vibration. These points are determined in part by the shape of the plate and frequency of the vibration.

STOPPING TIME
Phenomena too fast or too slow to be seen with the naked eye are explored in the Time section of *Take a Closer Look*.

OUR ENGINEERED WORLD

Thermal infrared imaging, first developed in the 1950s and 60s for military use, is now widely used in situations where detecting temperature variations would be advantageous. Examples include finding a person lost at night, maintaining a steady temperature while transporting food, detecting moisture and heat loss in walls, or as a high-tech security device.

Investigate!

Labs are not required for scientific inquiry: all that's needed is a question and an idea of how to test it. Designed to look like a house, *Investigate!* features activities and experiments that encourage visitors to practice thinking like scientists. It includes a section on energy conservation at home, mystery tools, an ancient device invented by Archimedes for water conveyance, and a component where visitors can conduct one of Galileo's famous experiments about gravity. These activities reveal the importance of asking the right questions and testing one's theories. Similar themes are explored in *The Light House*, *Science in the Park*, and *Seeing Is Deceiving*—adjacent exhibit areas where visitors can experiment in a wide range of subjects, including light and optics, forces and motion, and visual illusions.

IF A 100-POUND CANNONBALL AND A HALF-POUND MUSKET BALL ARE DROPPED AT THE SAME TIME, DO THEY FALL AT THE SAME SPEED?

Yes. Galileo proved that the speed of an object falling is independent of its mass. The force of gravitational pull on the objects is the same.

CREATIVE EXPERIMENTS
In *Investigate!*, visitors are encouraged to develop a hypothesis and create their own experiments. They can then test this theory and check their results to see if their guess was correct. The mannequin above is demonstrating an experiment by Galileo.

GOT WATER?
This water screw, left, is a device for "pumping" water and is attributed to Archimedes (287 to 212 BCE).

GUESSING GAME
Once data has been collected and the situation observed, visitors can make an educated guess.

UNDER THE SEA
Visitors can interact with an array of video screens in the *Virtual FishTank*. Backed by more than a dozen computers, the video screens create a virtual undersea environment.

Making Models

Models are best known as representatives for physical objects, but they also help scientists to conceptualize phenomena, systems, processes, and abstractions. *Making Models* features a replica of a giant grasshopper, nine different models of a human heart, and software programs that demonstrate other ways models are conceived and used. The exhibit includes the *Virtual FishTank*—a computer-generated undersea ecosystem populated by brightly colored, animated fish where visitors can alter the ecosystem and see the results.

More than just navigation tools, maps can also represent socioeconomic differences, reveal how diseases are spread, and show how plants and animals organize into food chains. *Mapping the World Around Us* features a large three-dimensional model of Mount Everest based on founding director Brad Washburn's survey work. The exhibit includes a reproduction of the earliest known map, a stick map used for centuries by Pacific Islanders, various population maps, and digital maps of complex structures such as the human genome.

MILESTONES
In 1937, Amelia Earhart asked Museum founding director and expert cartographer Brad Washburn to be the navigator for what was to be her ill-fated around-the-world flight. Washburn declined.

CLIMBING HIGH
This Mount Everest model, top, is based on the research of renowned mapmaker Bradford Washburn, founding director of the Museum of Science. Everest, the world's highest mountain (27,028 feet or 8,848 meters) is situated on the border of Nepal and China. It was first scaled successfully in 1953 by Edmund Hillary and Tenzing Norgay.

SOMETHING'S FISHY
Visitors set the rules for fish behavior in the *Virtual FishTank*, launch their digital creations into the ocean, and then watch what happens as the fish interact.

Natural Mysteries

Scientists use classification systems to make sense of the world's complex nature. By grouping observable similarities, nature's hidden patterns and meanings often reveal themselves in surprising ways. With a reference library of interesting objects, mystery environments, hands-on activities, and thousands of nature specimens available for examination, *Natural Mysteries* reveals how basic classification skills can change a scientist's perspective.

A Bird's World builds on this concept by illustrating the beauty and complexity of classification systems. The Bird Dictionary includes a specimen of every bird found in New England and a description of its particular features and habitats. Dioramas and interactive exhibits offer insight into bird behavior, language, and environment.

CLASSIFICATION CLUES
By examining the nails in the wall, items in the school room, and the tree roots at the foundation, visitors learn how to place a date on the Mystery School House in the *Natural Mysteries* exhibit, above.

PERSONAL DISPLAY
Visitors can sort and display a vast assortment of collected items at their disposal in Make Your Own Museum.

FOR THE BIRDS
The Bird Dictionary provides examples of every bird indigenous to New England, many collected over 100 years ago when conservation efforts didn't exist. Endangered species are highlighted, as are extinct species.

OUR NATURAL WORLD

One of the birds in the Museum's collection was originally and commonly known as a sparrow hawk. It was reclassified in 1983 and renamed the American kestral.

SMALL BUT FIERCE
Leopards are the smallest of the big cats, which include lions, tigers, and jaguars.

WHAT'S THE DIFFERENCE BETWEEN A LEOPARD, A JAGUAR, AND A CHEETAH?

Although all three cats are spotted, they live on different continents. Jaguars (the largest of these) live in the Americas, while leopards live in Africa and Asia. The slender cheetah, a native of Africa, is the world's fastest land mammal and has distinctive black facial stripes that drop down from both eyes.

Understanding the Cosmos

Even before Apollo 11 landed the first people on the Moon in 1969, humans have been captivated by outer space. Various exhibits throughout the Museum focus on space exploration and astronomy. The Charles Hayden Planetarium has taken visitors into the universe since 1958 by creating an immersive environment of cosmic phenomena projected on its great dome. In 2010, the Planetarium theater was renovated to include a digital full-dome projection system that allows the Museum to take audiences anywhere in the universe. This state-of-the-art equipment also enables the Museum to produce a new generation of immersive planetarium shows.

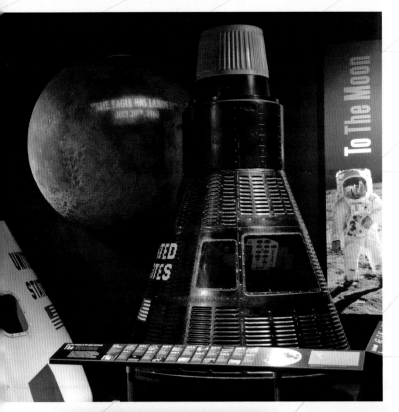

Cosmic Light, near the Planetarium entrance, includes a 12-foot-diameter model of the Sun and models of the planets sized in proportion to it. In other areas of the Museum, visitors can explore full-scale models of an Apollo capsule, a Mercury capsule, and a simulated control panel from the Lunar Excursion Module that landed on the Moon.

The Gilliland Observatory, located on the top level of the parking garage, reveals the universe outside the building's walls. Equipped with a computer-controlled 11-inch Schmidt-Cassegrain telescope, the observatory offers a dramatic view of stars, planets, and other deep space phenomena.

Confined Space
While the rockets that launched astronauts and cosmonauts into space were repurposed military rockets, the spacecraft they rode in were custom-designed. Weight was the primary concern, so everything had to be as small as possible.

Our Natural World
Planets and asteroids are held by gravity to the Sun, and they move in space without friction. Each naturally falls into an orbit that balances its own energy.

True to Scale
In the Planetarium lobby, there are bronze planet models in scale with each other and a 12-foot-diameter Sun. Saturn is tilted at 23 degrees relative to its orbit, similar to Earth.

IN THE STARS
Inside the Planetarium, audiences travel anywhere in the universe, at any point in time—past, present, or future.

OUR ENGINEERED WORLD

The Planetarium's state-of-the-art digital star projection system uses the most current scientific data to enable audiences to see the universe from a point other than Earth. Before this technology was available, audiences could only view the night sky as seen from our planet.

Understanding the Cosmos

IMPRESSIVE FEAT
Visitors can compare the International Space Station (ISS) model to models of earlier stations, including Skylab, Mir, and Salyut 1. The actual ISS is larger than a football field, assembled entirely in space by over 100 different crews with components from more than a dozen countries.

MARS AND BEYOND
The Mars Rover model in *Cahners ComputerPlace* can be controlled remotely to explore a simulated planet surface. The landing of the robot "Curiosity" on Mars in August 2012 added another tool for scientists to use as they probe for possible life elsewhere in the universe.

OUR ENGINEERED WORLD
The incredible photos of deep space that grace the Planetarium lobby were made possible by the revolutionary technology of the Hubble Space Telescope.

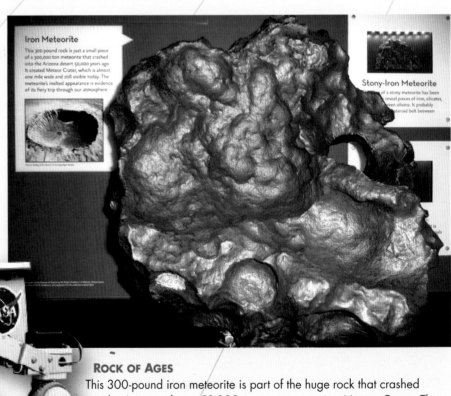

ROCK OF AGES
This 300-pound iron meteorite is part of the huge rock that crashed into the Arizona desert 50,000 years ago, creating Meteor Crater. The molten-looking surface of the meteorite shows the results of its fiery entry into the Earth's atmosphere.

GILLILAND OBSERVATORY
Weather permitting, Museum staff members host Friday night stargazing at the Gilliland Observatory, located on the top level of the parking garage.

SCOPING THE STARS
This 11-inch Schmidt-Cassegrain telescope is located inside the Gilliland Observatory. Visitors view their favorite objects in space through this telescope. The most requested objects to see are Jupiter, Saturn, and the Orion Nebula.

OUR ENGINEERED WORLD

Salyut ("Salute") 1 was the world's first space station, launched in 1971 by the Soviet Union. Now, more than 40 years later, the International Space Station is staffed by scientists from many countries working together on shared research.

BIG BANG
Every year during an event called EurekaFest, high school teams compete at the Museum to design, build, and test a wind-powered device capable of lifting a trash can three stories. The finale is the "big bang," when the cans are simultaneously dropped from their achieved heights.

Technology, Engineering, and Math

Scientists investigate the natural world, while engineers create the designed world, developing useful products and processes for society. These technologies, which are informed by science and mathematics, change human culture. The Museum of Science is dedicated to helping people better understand how engineering and technology relate to their daily lives. In 2004, it launched the National Center for Technological Literacy to enhance understanding of engineering and technology for people of all ages and backgrounds. This initiative focuses on putting engineering-based curricula in K – 12 schools and encourages science-related exhibits in museums.

In addition, various exhibits explore the concepts of engineering and technology. The *Gordon Current Science & Technology Center* showcases the latest scientific and technological advancements, while *Innovative Engineers* features several inventors whose ideas changed the world. Visitors learn how to put these ideas into practice in the *Engineering Design Workshop*. Technologies and methods for utilizing renewable energy are explored in *Energized!* and *Catching the Wind*. Finally, the field of mathematics—which is key to understanding engineering principles—is highlighted in the *Math Moves!* and *Mathematica* exhibits.

INSIDE VIEW
Beyond the X-Ray examines the world of medical imaging, from x-rays and Magnetic Resonance Imaging (MRI), to Computed Axial Tomography (CT) and ultrasound. These technologies allow doctors to look inside the human body without invasive surgery.

MATHEMATICAL BEAUTY
This chambered nautilus shell structure follows a mathematical logarithmic spiral pattern as described by 17th-century mathematician Jakob Bernoulli.

OUR NATURAL WORLD

Many repeating patterns in nature—fern leaves, branching rivers, seashell spirals, and even electrical discharges—are actually made up of complicated geometric shapes that repeat themselves on progressively smaller scales. These beautiful and intricate patterns are known as fractals.

Gordon Current Science & Technology Center

On the *Gordon Current Science & Technology Center* stage, Museum staff members give daily presentations on the latest scientific and technological advancements, and guest speakers present their newest research to the public. The Center also produces media on current science topics for the Web.

The products of human engineering are everywhere, from bridges to blow-dryers, smartphones to plastic wrap. *Innovative Engineers* highlights some of the engineering leaders who have solved everyday challenges. Biographical sketches reveal how the paths to engineering success are as diverse as the remarkable inventions these engineers have developed.

IN THE NEWS
Daily presentations on the Stage, top, demystify recent science developments that are in the news.

DISTINGUISHED SPEAKERS
Special presentations on the Stage have featured scientists, astronauts, and researchers who are directly involved in today's discoveries.

OUR ENGINEERED WORLD
PackBot robots, invented by engineers at iRobot Corporation, are used by the military to search for bombs. The company is also known for bringing robotics into households everywhere with the popular Roomba home vacuum cleaner.

EVERYDAY SCIENCE
Innovative Engineers features individuals who created groundbreaking products such as Kevlar, used in bullet-proof vests; mobile CT scanners; and water purification systems.

Engineering Design Workshop

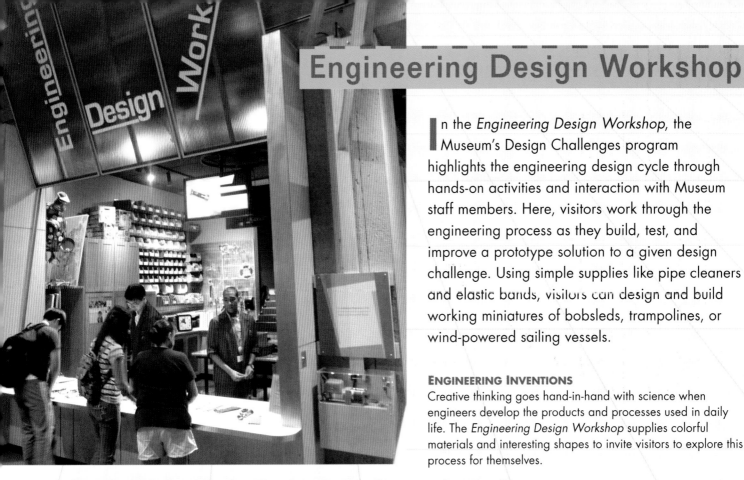

In the *Engineering Design Workshop*, the Museum's Design Challenges program highlights the engineering design cycle through hands-on activities and interaction with Museum staff members. Here, visitors work through the engineering process as they build, test, and improve a prototype solution to a given design challenge. Using simple supplies like pipe cleaners and elastic bands, visitors can design and build working miniatures of bobsleds, trampolines, or wind-powered sailing vessels.

ENGINEERING INVENTIONS
Creative thinking goes hand-in-hand with science when engineers develop the products and processes used in daily life. The *Engineering Design Workshop* supplies colorful materials and interesting shapes to invite visitors to explore this process for themselves.

VISITOR CHALLENGE
Staff-facilitated activities let visitors think like engineers as they work on solving the design challenge of the day.

OUR ENGINEERED WORLD
Boston inventor Edwin Land, creator of instant photography and founder of Polaroid Corporation, held over 500 patents. His instant cameras were first sold in 1948. They continued to be sold for 50 years until the advent of digital camera technology.

Sustainable Energy

In *Catching the Wind*, visitors learn how turbines transform wind power into electricity, what factors need to be considered when selecting and siting a turbine, and what trade-offs are made when choosing an energy source. Accounts of the decision process for turbine placement at several Massachusetts sites are described. The Museum's own Wind Lab—which features five styles of small turbines on the roof of the building—is also examined as a case study. The electricity produced by these turbines is tracked in real time, displayed, and compared with historical data.

While wind is a powerful source of energy, sunlight is the world's largest energy resource. More energy in the form of sunlight reaches Earth every hour than humans consume in a year. In *Energized!*, visitors learn how to make the best use of this energy and explore other possible energy sources for the future. Local innovations in green energy technology are also highlighted.

WIND POWER
There are five styles of small residential wind turbines installed on the Museum's roof. They can generate about 100,000 kilowatt hours of electricity annually.

SOLAR SAVINGS
The equivalent of 2.5 households worth of power generated by the photovoltaic array on the Museum's roof offsets some of the building's energy costs.

HOW DO SOLAR COLLECTORS WORK?

In contrast to conventional power plants, which use fuel such as coal, petroleum, or uranium to create heat, solar collectors use the Sun as a fuel source. Solar collectors use mirrors to focus sunlight at a central point, generating enough heat to boil water and then create electricity.

Mathematica

Developed in the 1960s by famous designers Charles and Ray Eames, *Mathematica* highlights the beauty and wonder of mathematics. Throughout the exhibit, which is designed in classic modernist style, there are compelling images and stories from many branches of mathematics, including probability, topology, Boolean algebra, geometry, calculus, and logic. A timeline of these achievements is displayed on the wall, and a selection of images reveals mathematical patterns in nature, such as the seeds in a sunflower and the Golden Spiral in the shell of a chambered nautilus. Other displays show soap bubbles forming on wire shapes, revealing the minimal surface area for that shape, and a train car running along a giant, endless Mœbius band.

MILESTONES
Mathematica was displayed in the IBM Pavilion at the 1964–65 New York World's Fair.

VISUAL MATH
A helix, close-packed spheres, and dozens of other math-based models demonstrate mathematical principles.

Math Moves!

Understanding ratios and proportions is a necessary part of daily life. *Math Moves!* brings these mathematical concepts to life, using shadows, shapes, motion, and sound to do so. The exhibit reveals how the human body can become a measuring tool by inviting visitors to climb first into a huge chair and then a tiny one to compare size relationships. A shadow table shows the visual connection between size of an object and its distance from the observer.

IN THE SHADOWS
The Shadow Fractions activity in *Math Moves!* gives visitors a fresh perspective on the size and shapes of shadows relative to viewpoint and distance from light source.

LEARNING BLOCKS
Scaling blocks demonstrate the relationship of volumetric to linear dimensions.

SUPER CHARGED
The power of lightning is explored in dramatic daily demonstrations of the Museum's Van de Graaff generator.

OUR NATURAL WORLD

When lightning hits the beach, the intense heat can fuse the sand into a tube of glass called a fulgurite.

Thomson Theater of Electricity

Named after inventor and scientist Elihu Thomson, the Thomson Theater of Electricity is home to the world's largest existing air-insulated Van de Graaff generator. The generator was designed and built in the 1930s by Dr. Robert J. Van de Graaff as an atom smasher and particle accelerator for the Massachusetts Institute of Technology (MIT). MIT donated and moved the generator to Science Park in 1955, and in 1980 the Museum completed the building to house the three-story apparatus. Demonstrations run daily, with lightning bolts reaching a million volts or more and stretching 15 to 20 feet. The show features Tesla coils, including one with sparks that make music!

The Theater also houses interactive exhibits such as an electrical device that can propel an aluminum disc 20 feet in the air and another in which glowing swirls made by electrical energy are attracted to the visitor's touch. Artifacts from Thomson's lifelong work with electricity are on display. In *WeatherWise*, visitors explore weather on a global, national, regional, and local scale in order to develop skills to make their own short-term forecasts. Displays range from a miniature touchable tornado to a globe projecting current weather systems around the world in real time.

TOUCH AND GLOW
Swirling ionized gases dance to the proximity of a visitor's touch in this plasma tube built by inventor Bill Parker.

MUSICAL ELECTRICITY
The Museum's musical Tesla coil plays recognizable musical selections, from symphonic to movie themes.

DOES THUNDER ALWAYS ACCOMPANY LIGHTNING?

Yes. Lightning superheats surrounding air molecules, producing a shock wave that expands quickly and generates the sound of thunder. If a person sees lightning but doesn't hear the clap of thunder, it simply means that the lightning is very far away.

Discovery Center

A miniature museum aimed at young visitors from infants to eight year olds and their caregivers, the *Discovery Center* offers hands-on activities that are designed to encourage discovery through play. The Center emphasizes the use of touchable objects and includes an extensive collection of artifacts for children to explore natural history, physical sciences, and technology. Here, young scientists can assemble an animal skeleton, touch a real fossil, or observe a variety of live animals up close. The Experiment Station offers a rotating schedule of science experiments and engineering challenges, and the Discovery Boxes allow an in-depth exploration of specific science topics. The Bee Hive, Robin's Nest, and Geology Field Station are simulated environments that encourage learning during play. At the Bee Hive, children five years and younger turn into honeybees—looking for flowers, dancing to communicate with other bees, and returning to the hive with honey.

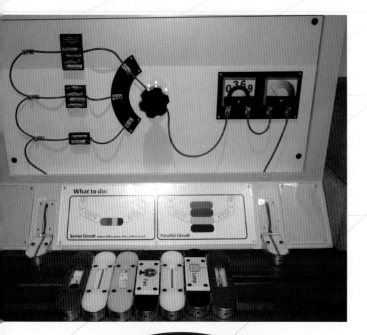

ELECTRIC EDUCATION
Children can play with simple electrical circuits using magnetic component building blocks in the *Discovery Center*.

WET AND WILD
Young visitors are drawn to the water play area, above left, where they can use water to explore and experiment as they develop their early scientific thinking skills.

DIVING DISCOVERIES
The Discovery Boat gives young visitors the opportunity to be marine biologists and to discover underwater wonders such as seashells, coral, horseshoe crabs, and whelk egg cases.

HANDS-ON LEARNING
Young visitors to the *Discovery Center* can observe live animals up close and have hands-on experiences with hundreds of natural history specimens.

OUR NATURAL WORLD

A robin's nest is built primarily by the female. In the spring, she and her mate gather grass, small sticks, and any string-like materials they can find, and in less than a week, she weaves them together to build the nest. She usually lays about three to four greenish-blue eggs. Both parents protect the nest and feed their young.

A BIT OF WHIMSY
Archimedean Excogitation by George Rhoads mesmerizes with its whimsical mechanics.

Art in the Museum

Science and art may appear to be worlds apart, but both disciplines interpret, study, and explore the mysteries of the natural world. The Museum has several exhibits that investigate the marriage of science and art. In the Art & Science Gallery, temporary exhibits of artwork explore science and technology. Shows have included photographic illustrations of how people affect the environment, the art and science of bonsai trees, and cultural differences in how humans approach and consume food.

The Museum is also home to permanent science-and-art exhibits. In *Archimedean Excogitation*, balls race around tracks, bounce down steps, hit chimes, spin wheels, and swing pendulums. In the Atrium, *Soundstair* is an example of participatory and responsive architecture, which creates music that is unique to each visitor's step. The Atrium also features *Polage*, a polarized light collage that depicts images that link people to one another, to history, to the planet, and to the universe. In the main lobby, *Capturing the Arc* is an interactive light sculpture consisting of 27 curved acrylic panels, each lit by computer-programmed LEDs.

THE ART OF SCIENCE
The Museum's Art & Science Gallery offers a contemplative place to view art that has a basis in, or connection to, science content.

COLORFUL CREATION
In *Capturing the Arc* by Joey Nicotera, the vibrantly colored light seen in each panel passes gradually from one color to the next, producing a continually changing display.

MILESTONES

The Museum opened its new Art & Science Gallery in 2009. The connection between these disciplines has always been part of the Museum's culture. Between 1973 and 1985, the Museum hosted 38 art shows ranging from bronze animal sculptures and pastel drawings, to photography and cartoon animations.

Art in the Museum

TINY TREES
This tropical dwarf jade plant is approximately 35 years old. It is featured in *Bonsai: Creating Art Through Nature*, an exhibit that was created in partnership with Bonsai West.

WHAT IS AN LED?

An LED, or light emitting diode, produces light from semiconductor chips—and not from heated filaments. LEDs generate far less heat and use far less electricity than their older light bulb counterparts.

ANIMAL ART
Katharine Lane Weems's animal sculptures, mainly in bronze, reveal her careful observation of anatomy and the integration of art with science. Weems became one of the most recognized animal sculptors of her time, breaking away from early 20th-century social standards for women during her 70-year career.

ART OF ILLUSION
Interactive displays, such as this drawing activity using spherical mirrors, help explain the illusion artwork created by graphic artist M. C. Escher.

KALEIDOSCOPE OF LIGHT
Polage, created by Austine Wood Camarow, stretches across a large portion of the Atrium wall. It consists of 38 hexagonal light boxes displaying images that are created with layers of polarized film and cellophane. When seen through rotating polarized viewers, the pictures on the wall change, hidden objects and colors appear, and images seem to move.

OUR ENGINEERED WORLD

Edwin Land developed a low-cost method of making polarizing filters. Polarizing filters can be used to filter out glare, create 3-D illusions, and produce colorful artistic effects with light as seen in *Polage*, the Museum's polarized light collage. Polarizing filters are also used in LCD displays.

BIG GAME
The *Colby Room* represents a time in the early 20th century when big game hunters like Teddy Roosevelt were admired for their daring adventures in the wild.

Special Collections

Amid the Museum's dinosaur fossils, lightning bolts, and virtual technology are several unique collections. *Then and Now* chronicles the Museum's development over the last 180 years, showcasing natural history specimens in classic 19th-century style, technology from the mid-20th century, and some of the Museum's first interactive exhibits. Visitors can also find architectural renderings of future plans and of an exhibit on computer animation.

The *Colby Room*, which opened in 1965, is another portal to the past. The room recreates adventurer and Museum benefactor Colonel Francis T. Colby's den in Hamilton, Massachusetts. Pelts, mounted heads, horns, and antlers are scattered throughout the room, and East African statues, ivory figurines, and Ethiopian artwork adorn the walls and tables.

At the entrance to the Educator Resource Center is a special collection of 30 bronze animal sculptures by Boston-born artist Katharine Lane Weems. By carefully observing animal anatomy and behavior, Weems brought her subjects to life with incredible biological accuracy. The Museum's sculptures represent the largest Weems collection in the world.

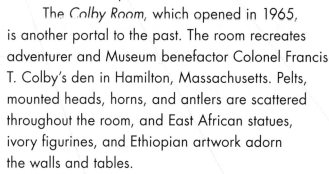

OUR ENGINEERED WORLD
The first electronic office calculators from the late 1960s could barely fit on a large desk. With a 512-bit magnetic core memory and glowing Nixie tube display, even the simplest calculation took several seconds.

FAMOUS CONTRIBUTORS
Since the Museum's start in 1830, many well-known scientists have contributed to its success. John James Audubon, for example, a noted international naturalist and ornithologist, contributed to the Museum's collections, and studied its collections for his famous illustrations.

MUSEUM EVOLUTION
The Museum of Science has undergone many changes over the years, from its original goal of collecting natural history specimens to its current mission of helping visitors understand both the natural and engineered worlds.

Temporary and Traveling Exhibits

The Museum of Science has a long history of presenting exceptional temporary and traveling exhibitions. In 1985, the Museum hosted its first blockbuster exhibition, *China: 7,000 Years of Discovery*. The exhibit included more than 300 artifacts and objects, along with over a dozen artisans from China who demonstrated traditional arts, crafts, and technologies ranging from papermaking to silk brocade weaving.

With the success of *China*, the Museum decided to host large-scale exhibitions on a regular basis. Since then, other cultural world heritage offerings have included *Ramesses the Great* (1988), *Leonardo da Vinci* (1997), and *A Day in Pompeii* (2011).

Science and engineering exhibitions have included *Soviet Space* (1990), *BodyWorlds 2* (2006), and *Star Wars: Where Science Meets Imagination* (2005), which was developed by the Museum and premiered in Boston before traveling to museums worldwide. Other exhibits, such as *The Dinosaurs of Jurassic Park* (1993), *Star Trek* (1992), and *The Lord of the Rings* (2004), have touched on popular culture, incorporating paleontology, future technology, and motion-capture animation in fun, immersive environments. The Museum also collaborates with other museums nationwide to develop and tour a wide range of exhibits.

WINGED CONTRAPTION
This model of Leonardo da Vinci's ornithopter—a flying machine that he designed after studying the flight of birds—was displayed at the Museum.

INTERNATIONAL COOPERATION
In 1990, the Museum hosted an exhibit on the history and achievements of the Soviet Space program, which included more than 50 spacecraft and artifacts. Developed by the Museum of Science in cooperation with Glavkosmos, the Soviet Space agency, the exhibit was the first of its kind to travel to the United States.

STANDING GUARD
Terra cotta warriors protected the burial site of the first emperor of China. This full-sized figurine and horse were included in the exhibition *China: 7,000 Years of Discovery*.

HEAVY LIFTING
In conjunction with the 1988 exhibition *Ramesses the Great*, the Museum built an Egyptian temple on its front plaza that housed a 27-foot-tall, 60-ton granite statue of Ramesses, the Colossus of Memphis.

Our Natural World

The eruption of Mount Vesuvius in AD 79 created a pyroclastic flow where volcanic rock and ash that had been thrown a mile into the atmosphere fell to Earth, flowing at incredible speed and with enormous heat to the city of Pompeii. The population was caught off guard; in fact, the people of Pompeii didn't even have a word for volcano in their vocabulary.

CAPTURING TRAGEDY
Ash from Mount Vesuvius buried the unfortunate inhabitants of Pompeii, Italy, in AD 79, creating molds of their remains that would lie undiscovered for nearly two millennia.

Temporary and Traveling Exhibits

ANCIENT EGYPT
In 2002–2003, *Quest for Immortality* was the second major Egyptian artifact exhibit shown by the Museum.

MILESTONES

In 1984, the Museum of Science teamed up with six other science museums to form the Science Museum Exhibit Collaborative to develop and circulate top quality science-related temporary exhibits. *Star Wars: Where Science Meets Imagination* was one of many successful results.

IN A GALAXY FAR, FAR AWAY
Star Wars: Where Science Meets Imagination has toured 20 cities throughout the United States and Australia. C-3PO (costumed actor Anthony Daniels) was among the illustrious guests who attended the Museum's 2005 opening of the show.

Reaching the Future

The Museum of Science is committed to bringing students and teachers into the cutting-edge world of science and technology. To do this, the Museum focuses on several initiatives. The Computer Clubhouse uses technology to enrich the lives of inner-city youth. Here, young people work with adult and teen mentors on projects they choose, creating art, music, video, scientific simulations, animations, kinetic sculptures, robots, and Web pages. The Museum houses the flagship Clubhouse, which develops and tests new projects and activities before rolling them out to other sites, and the Intel Computer Clubhouse Network, which provides support to Clubhouses worldwide.

For educators, the Museum has created the Educator Resource Center (ERC) in the Lyman Library. Teachers can receive a range of support at the ERC, from half-day guidance for field trip planning to intensive five-day science education programming and professional development workshops. In the ERC's 18,000-volume collection, educators can find science-fair project ideas, inquiry-based approaches to teaching, hands-on classroom activities, popular science DVDs, and a variety of performance assessments.

HIGH TECH LEARNING
In the Computer Clubhouse, boys and girls are encouraged to explore and develop their interests in design, science, and engineering.

AN ENRICHING EXPERIENCE
The Educator Resource Center offers a variety of opportunities for teacher enrichment.

ROLE MODELS
Adult mentors serve as role models for the children, creating a rich and diverse learning community.

TECHNOLOGY TOOLS
Computer Clubhouse members use professional technology tools to develop computer-based projects inspired by their own ideas, gaining valuable job and life skills in the process.

MILESTONES
The Computer Clubhouse is a project of the Museum of Science in collaboration with the MIT Media Laboratory. Since 1993, more than 100 Computer Clubhouses have been established in 20 countries, reaching over 20,000 youths annually.

OUR ENGINEERED WORLD

A new approach developed by scientists for targeting viruses may help cure the common cold and more! Engineered molecules attach to the virus and activate the self-destruction of the host cell, stopping the infection in its tracks.

HALL OF HUMAN LIFE
Pictured is an artist's rendering of the entrance portal to the *Hall of Human Life*, where visitors will pass through a giant cell membrane to an evolving, dynamic, participatory experience.

Looking Ahead

The Museum is currently undergoing a significant redesign of its galleries to highlight the interdependency between the natural world and the world as changed by human design. A cohesive story will unfold, revealing how science unravels the mysteries of our natural world, while engineering devises the means that better enable us to live and thrive in it.

The *What Is Technology?* exhibit will focus on the characteristically human way we meet the challenges of nature by designing systems, processes, and technologies that enable us to thrive in our world. The *Yawkey Gallery on the Charles River* will make use of the Museum's unique position on the Charles River to illustrate the relationship between the natural environment and human technologies. This gallery will address themes in ecology, earth systems, biodiversity, technology, and human impact. The *Hall of Human Life* will explore modern biology and its implications for our health from evolutionary, environmental, cellular, and microbiological perspectives. With advanced technologies and inventive programming, Museum visitors will explore how we, as individuals and as societal groups, adapt to and change our environment, which in turn affects our health and our long-term well-being.

The natural world and the human-engineered world are inextricably linked. Changes in one affect the other, often in unforeseen ways. The Museum will afford visitors the opportunity to explore these connections between the natural and engineered worlds, better enabling them to make informed choices for themselves and for society.

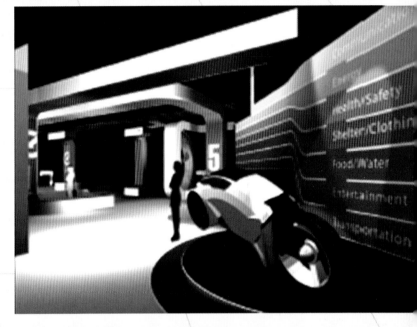

TRENDS IN TECHNOLOGY
The latest in technological innovation will be showcased in the *What Is Technology?* exhibit.

OUR NATURAL WORLD

Once an estuary with mud flats and salt marshes affected by daily tides, the Charles River Basin today is a wildlife habitat for hundreds of animal and plant species, each playing a role in the ecology of the region.

CONNECTING OUR WORLDS
Entry to the Exhibit Halls will reflect the Museum's overall vision of the connections between the natural world and our technological world.

Yawkey Gallery on the Charles River

The *Yawkey Gallery on the Charles River* will immerse visitors in the setting of Boston's own urban watershed to examine the diversity of life sustained here and the integral role that humans play in this dynamic ecosystem. This gallery will tell the story of the river basin and its life as both a natural-world habitat and an engineered environment, themes that will be echoed throughout the Museum. The exhibit will examine the complex ecosystems in and around the river and the human activity that has shaped and altered this landscape for thousands of years.

GLIMPSE OF NATURE
The three-story windows overlooking the Charles River will provide a vantage point to learn about biodiversity and ecosystems.

What Is Technology?

The *What Is Technology?* gallery will seek to broaden the public's definition of technology, reaching beyond consumer electronics to include every human-made aspect of our daily lives. Technology encompasses our entire engineered world—not only items we create, but also the materials and processes we use to create them. Focusing on areas ranging from food and transportation to energy and entertainment, the exhibit will feature frequent updates to keep it current in our rapidly evolving technological world.

SCIENCE IN ACTION
State-of-the-art interactive components will create a dynamic visitor experience in the *What Is Technology?* gallery.

Hall of Human Life

The *Hall of Human Life*, opening in fall 2013, will revolutionize the way the public engages with human biology, helping visitors to understand their own bodies and make informed choices as advances accelerate in health care. The exhibit focuses on the ways physiological change affects individuals on a daily basis just as evolutionary change affects human populations over time. In the context of five distinct environments—food, physical forces, non-human organisms, social interaction, and time—visitors will explore how these environments change us, how we change them, and how those changes can impact the future of our species.

Using state-of-the-art technology, visitors will track their progress at each station with a barcoded wristband, measuring how their biology responds to dynamic environments, and then make comparisons with the larger population as data is assimilated and analyzed. The exhibit will also engage the public with provocative questions in subjects ranging from genetics and new medical technologies to exercise and nutrition.

SCREEN TIME
The Human Body Theater will be an interactive experience featuring large, panoramic screens and a 3-D projection onto a human anatomical form.

ENGAGING VISITORS
At the Exploration Hub, staff and volunteer interpreters will engage visitors with hands-on demonstrations and short experiments related to themes found throughout the *Hall of Human Life*.

MILESTONES

In 1975, the Museum opened a fully equipped and staffed Red Cross blood donor center in its Hall of Medical Science. The center operated for more than a decade and served as both an exhibit and a community service facility. The Museum was the first in the country to house such a facility.

ON THE WATERFRONT
The Museum of Science sits on a dam built over 100 years ago to stabilize daily tidal fluctuations in the Charles River Basin. From this prime location, visitors are treated to an unparalleled waterfront view of the Boston and Cambridge skylines.

A Note from the President

Always reinventing itself to meet society's changing needs, the Museum of Science has a remarkable story to tell: over 700 interactive exhibits, leadership in technological literacy across the nation and the world, and daily presentations by our educators about discoveries and advances in science and engineering.

This is an extraordinary time for the Museum. To meet the challenges of the 21st century, the Museum has embarked on a significant and comprehensive transformation of its Exhibit Halls. A focus on biodiversity and biological change coupled with increased concentration on technology and engineering will create an opportunity for the visitor to probe the delicate balance that exists between them.

Dynamic new galleries and compelling programs will invite visitors on a learning journey that highlights this profound link between our global ecosystem and the world we design, as each continuously alters the other in complex ways.

In focusing on the interdependency between our natural and engineered worlds, the Museum hopes to engender recognition that through our behavior and innovation, we literally engineer our future—and the future of our planet.

Ioannis (Yannis) N. Miaoulis
Museum director and president

Museum of Science

One of the world's largest science centers, the Museum of Science is known for its innovative programming and more than 700 interactive exhibits—a place where everyone can participate in the excitement of science and technology. The Museum also leads a national initiative to raise awareness and understanding of science, technology, engineering, and mathematics in schools and museums.

Museum of Science
1 Science Park
Boston, MA 02114
617-723-2500
mos.org

The *Museum of Science: Official Commemorative Guide* was written, compiled, and edited by Larry J. Ralph and Mary Ann Trulli.

Special thanks to the many staff members at the Museum of Science who contributed to the making of this guide.

Ioannis N. Miaoulis, *President and Director*
Larry J. Ralph, *Director, Education Enterprises;*
Project Manager
Carl Zukroff, *Director, Marketing Communications*

Photo Credits:
Unless otherwise indicated, all photos are copyright of the Museum of Science. ©Brilliantpictures.com: 7; Cambridge Seven Associates, Inc. (design renderings): 59a, 60b; ©Dave Desroches: 62; Michael Horvath (design renderings): 58, 59b, 60a, 61b; Beth Malandain (design rendering): 61a; ©Michael Malyszko: 14c, 16c, 24c, 27c, 32b, 33, 36, 39b, 39c, 41b, 44b, 56a, 63, back cover; Emily Marsh: 24a, 30b, 38c, 51b; Nate Marsh: 50, back cover; Larry J. Ralph: 8, 9a, 9b, 9c, 10a, 10b, 11, 13a, 13b, 14a, 15b, 16a, 17, 18a, 18b, 18c, 19, 20a, 20b, 20c, 21a, 21b, 21c, 22, 23b, 23c, 24b, 25, 26a, 26b, 26c, 27a, 27b, 29a, 29b, 30a, 30c, 31, 32a, 34a, 34b, 34c, 35a, 35b, 37a, 37b, 39a, 40a, 40c, 41a, 41c, 43a, 43b, 44a, 44c, 45, 46, 47a, 48a, 48c, 49, 51a, 52a, 52b, 52c, 53, 54, 55a, back cover; Eric Workman: 12, 14b, 15a, 15c, 16b, 23a, 40b, 42, 47b, 48b, 55b, 55c, 56c, back cover; David Rabkin: 2, 38a, 38b; ©stevemarselstudio.com: cover; Carl Zukroff: 4; Museum of Science historical archives: 5a, 5b, 6a, 6b, 6c; Star Wars images courtesy of Lucasfilm Ltd.: 55b, 55c; IMAX film image courtesy of MacGillivray Freeman Films: 7

BECKON BOOKS

The *Museum of Science: Official Commemorative Guide* was developed by Beckon Books in cooperation with the Museum of Science and Event Network. Beckon develops and publishes custom books for leading cultural attractions, corporations, and nonprofit organizations. Beckon Books is an imprint of Southwestern Publishing Group, Inc., 2451 Atrium Way, Nashville, TN 37214. Southwestern Publishing Group, Inc., is a wholly owned subsidiary of Southwestern, Inc., Nashville, Tennessee.

Christopher G. Capen, *President, Beckon Books*
Monika Stout, *Design/Production*
Betsy Holt, *Editor*
www.beckonbooks.com
877-311-0155

Event Network is the retail partner of the Museum of Science and is proud to benefit and support the Museum's mission to transform the nation's relationship with science and technology.
www.eventnetwork.com

Copyright © 2013 Museum of Science

All rights reserved. No part of this book may be reproduced or transmitted in any form or by any means, electronic or mechanical, including photocopying or recording, or by any information retrieval system, without the written permission of the copyright holder.

Star Wars™ & © 2013 Lucasfilm Ltd. All rights reserved. Used under authorization.
IMAX is a registered trademark of the IMAX Corporation.
Species Survival Plan (SSP) is a registered trademark of the Association of Zoos & Aquariums.
National Center for Technological Literacy, Computer Clubhouse, and the Virtual FishTank are registered trademarks of the Museum of Science.

ISBN: 978-1-935442-23-3
Printed in the United States of America
10 9 8 7 6 5 4 3 2 1